THE HEYDAY OF THE CLASSIC COACH

Ian ALLAN Publishing

KEVIN LANE

First published 1994
First reprinted 1996
This impression 1999

ISBN 0 7110 2270 4

© Ian Allan Publishing Ltd 1994

Published by Ian Allan Publishing

an imprint of Ian Allan Publishing Ltd, Terminal House, Shepperton, Surrey TW17 8AS.
Printed by Ian Allan Printing Ltd, Riverdene Business Park, Hersham, Surrey KT12 4RG.

Code: 9905/2

Front cover: This ex-Glentons centre-entrance AEC Reliance with Plaxton Panorama bodywork, which was new in 1965, was being operated by Warren's Coaches of Ticehurst, West Sussex, when photographed in 1973 at the old Grantham bus station. *John Jones*

Back cover: A trio of West Wight Motor Bus Co Bedford SBs is pictured at Yarmouth on the Isle of Wight. In the centre VVP911 is fitted with a butterfly-fronted Duple Vega body, whilst UDL453 and YYV488 have both received Duple Super Vega bodies. *Martin Llewellyn*

Title page: That classic of classics - the Bedford OB - first appeared in 1939, although the war interrupted production until 1945. During the war the Utility OWB took the OB's place. The OB could be found countrywide and, although it was bought by some major operators, it was more at home with the smaller independent. Although the last OBs were built in 1951 it continued to be a familiar sight well into the 1970s. Bedford OB owner and AA member MacLeod of Duntulm was still running Duple Vista-bodied SX7039 on his service from Kilmaluag when seen in Portree on the Isle of Skye in July 1970. *Chris Lodington*

Introduction

As far as coaches go, the terms 'heyday' and 'classic' are open to much interpretation. In this volume I have not really attempted to give my view, other than to limit coverage from 1945 to the late 1960s. As far as the vehicles themselves go, one man's classic may well be another's load of old junk. It is true that many may agree that the Bedford OB is a classic, there may be rather more debate over the Daimler Roadliner! The vehicles illustrated in this book reflect a wide field, with many of the types built during this period depicted. This was no easy task as many of the photographers who worked in colour in the 1950s and 1960s appeared largely to ignore the coach in favour of the service bus (I was, alas, equally to blame when I started to take bus pictures in the late 1970s). Perhaps, with the expense of slide films then, this was understandable, indeed such considerations still exist today.

The development of coaching since the war makes for interesting study. The relaxation of travel restrictions imposed by hostilities saw a huge expansion of express, limited stop, tour and excursion work and, with so few chassis available, it was a case of make do and mend for a couple of years. This meant extending the lives of already well-worn prewar vehicles by various means. The fitting of new bodies to old chassis spawned a whole new industry as new bodybuilders sprang up to compete for this lucrative business. Although this continued as new chassis became increasingly available, their numbers declined as the boom lessened and as the larger builders began to dominate the market.

As the 1950s progressed, so coach design developed also. From 1950, maximum vehicle length was increased to 30ft, allowing a higher seating capacity. It was also at this time that underfloor-engined coaches were coming on to the scene and this led to improved designs, which, coupled with more seats, saw the traditional half-cab off the road. The first generation of underfloor-engined coaches were heavy, however, and there was always a place for the front-engined lightweight, such as the trusty Bedford SB.

From 1961, maximum vehicle length was again increased, this time to 36ft. By now we were into the motorway age, the new network of roads opening up a whole new sphere of operation, shorter journey times offering a serious alternative to the railways. The vehicles themselves were becoming more sophisticated as were the facilities enjoyed by the passenger; better heating and ventilation, on-board refreshments and toilets. The tour and excursion business was also flourishing and private hire was also on the increase — often using mini-coaches, a useful addition to many fleets.

The coaching scene since the late 1960s has been increasingly dominated by European influence, so that my self-imposed cut-off date could then be said to be the end of a heyday. Looking through the book, it is sad to reflect that almost all of the chassis manufacturers and bodybuilders are now no more, even names like Bedford and Leyland that we thought would go on forever have been consigned to history. Many of the operators have also fallen by the wayside; no more Midland Red as such, Barton is now a fleetname for Trent, while the likes of Yelloway and Standerwick have gone altogether. I hope that the pictures that follow will ensure that they are not forgotten.

Thanks are due to the photographers who have made this book possible and to their foresight in recording the coaches in colour in the first place. Special thanks are extended for the wealth of extra information supplied to John Bennet and to John Hambley for allowing me access to his collection of PSV Circle fleet histories. Thank you also to my wife, Maureen, for allowing the dining room table to become engulfed in bus books, yet again.

Kevin Lane
May 1994

The first AEC single-deck chassis after World War 2 was the Regal I of 1946, which was similar to its prewar counterpart. It was replaced by the improved Regal III two years later. Of traditional front-engined design, the model sold well to vehicle-starved operators of the late 1940s and early 1950s, although eventually the advent of the underfloor-engined Regal IV in 1950 saw its demise by 1955.

LRA907 is a 1947 Regal III, which was new to Woolliscroft (Silver Service), Matlock, fitted with a Duple 'A' C35F body. It lasted an impressive 23 years with Woolliscroft before withdrawal and preservation in 1970. The coach is seen at the old Matlock bus station in 1967, with a Bedford SB of the same operator behind.
Martin Llewellyn

3

Left: Another AEC Regal III, this time with Scottish Aviation bodywork, is DAG 607 of Lennox on the Scottish island of Arran. The twin horns and stripped radiator grill certainly make for an eye-catching vehicle. *Alistair Douglas/Photobus*

Above: As will be seen later in the book, it was not uncommon for some bodybuilders to fit full-fronts to front-engined designs in order to give a more modern look. This Regal III of Bere Regis & District Motor Services carries Beccols bodywork, seating 31, and was new, along with

JP8147 in the fleet, to Smith (of Wigan) in 1950. Both passed to Bere Regis in 1955. Also in this large and varied fleet were JP8050/51, Beccols-bodied Leyland Tiger PS2s, which were also new to Smith in 1950.
Alistair Douglas/Photobus

As noted earlier, the first underfloor-engined chassis for AEC was the Regal IV, produced from 1950 until 1955, which proved popular with coach operators.

London Transport was the largest bus operator of the type with its 700-strong RF class and it also ran 15 of the RFW class, which were touring coaches with ECW bodywork built to the then maximum permitted length of 30ft and were 8ft in width. Delivered from May 1951, the class saw use on sightseeing tours and private hire work; here No RFW 13 is seen on a City Tour at Victoria in 1962. The class was withdrawn in 1964 and many, including No RFW 13, were exported for further service in Ceylon. *Iain MacGregor*

6

An interesting line up in the yard of Dodds (Troon) in 1964: OWA209 is
an AEC Regal IV with Windover Kingsway bodywork; LTJ904 is a
Bellhouse Hartwell-bodied Foden PVSC; while, on the right, is CAG39, a
Bristol L6G with a Burlingham body sporting a Plaxton full-front.
Iain MacGregor

Left: The problem with the first generation of underfloor-engined chassis was that they were heavy and consequently thirsty on fuel. From the Regal IV, AEC turned its attentions to a completely new model, the Reliance, which appeared from 1953 and became a highly popular basis for both bus and coach applications for many years. AEC merged with Leyland in 1962 and the gradual rationalisation of models was inevitable. The Reliance bowed out in 1979 with the closure of the Southall works - a long run indeed.

East Kent was a good AEC customer, buying its first Reliances in 1955 and then obtaining more regularly for the next 20 years. MJG47 was one of a dozen supplied with Beadle C32C bodies in 1957. It was used for touring work until 1970. The entire batch was withdrawn in 1975/76, with MJG47 going for scrap. The coach is seen at Margate Harbour in 1971. *Richard Mellor*

Below left: DCU23 was a 1960 AEC Reliance fitted with attractive Duple Britannia C43F bodywork and operated by Hall Bros of South Shields. It is seen in Doncaster when brand-new working between Coventry and Newcastle. Its life was short, as it was destroyed, along with seven other vehicles, in a fire at South Shields in October 1962. Hall Bros was taken over by Barton in 1967. *Roger Holmes*

Right: Sitting in the sun at Huntington Street bus station, Nottingham, in September 1967 is a Black & White AEC Reliance/Duple C37C, No A221, on layover before returning to Cheltenham. The service was part of the Associated Motorways network, in which six operators were involved: United Counties; Red & White; Royal Blue; Midland Red; Bristol; and, of course, Black & White. In later years these were joined by Lincolnshire Road Car, Crosville and Eastern Counties and the network covered most of the country. This pool was terminated from October 1973, with the take over by National Travel. *G. H. F. Atkins*

The 36ft Reliance became available following the increased maximum permitted length in 1961. Premier Travel of Cambridge was a regular customer of the Reliance, both new and secondhand. A number carried the unusual (for an English operator) Alexander 'Y'-type bodywork. No 180, DCE800C, was one of a pair bought in 1965 and is seen bound for Blackpool in 1967 next to a Midland Red D9 in Leicester. *Mike A. Sutcliffe*

Left: From 1962, London Transport placed coach versions of the Routemaster class on to Green Line services, following on from the prototype (No RMC 4) of 1957. A longer, 30ft, version, the RCL, went into Green Line service from 1965. Seating 65, the RCLs were fitted with platform doors, fluorescent lights and luggage racks. The RMC and RCL classes passed to London Country Bus Services in 1970 to continue on Green Line duties, but most had been bought back by London Transport between 1977 and 1980. Initially put to work as training buses, all of the RCLs (except No RCL 2221) were returned to passenger duties, following certain modifications, in which role they remained in service until withdrawal between 1983 and 1985. Some were sold, but 11 were transferred to London Coaches for sightseeing and private hire work.

One of those 11 was No RCL 2260, which is pictured leaving Sevenoaks for Windsor on Green Line route No 704, a journey that passed through the heart of the capital and took almost three hours in theory (although with traffic congestion it probably took longer). No 6066, a 1963 Maidstone & District Daimler Fleetline/Northern Counties, heads for home alongside, in this March 1969 view. *John Aldridge*

Left: The Albion Victor FT, a lightweight forward-control chassis, appeared from 1947. The initial model was the FT3AB. The type developed throughout the 1950s until it was replaced by the Victor VT range in 1959.

Two FT3ABs are pictured with Chiltern Queens, Woodcote, Oxfordshire. NPA461/2, both new to Whites (Camberley) in 1949, carried Wadham C31F bodies. Sandwiched in between them is another ex-Whites vehicle, NPE61, a Wadham-bodied Guy Vixen, new in 1949. Chiltern Queens remains an interesting fleet, although the variety to be found during the 1950s and 1960s is but a memory.
Alistair Douglas/Photobus

Above: In 1937 Albion launched the Valkyrie CX chassis for single-deck bodywork. Production was interrupted by the war, but resumed in 1945 to continue until 1950 when the Valiant CX39, introduced in 1948, took over. Most Valiants received coach bodywork, typified here by Red & White No S4050 (HWO362), a 1950 delivery with a Duple C33F body. Red & White was an enthusiastic Albion customer in its independent days, but Bristols became the standard following acquisition by the British Transport Commission in 1950.

No S4050 had started life as No 62, being renumbered C2350 in 1951, DS4050 in 1959 and finally S4050 in 1962 - all reflecting its gradual reduction in status. The last two digits denote year of manufacture. Withdrawal came in 1963, with the vehicle passing to a Chepstow dealer. *Mike A. Sutcliffe*

Right: Albion entered the underfloor market in 1951 with the KP71NW, which was something of a false start as it turned out since only two were built. The tiny Nimbus followed in 1955, but the first full-sized single-deck chassis was the Aberdonian MR11L, which was produced from 1957 until 1960.

Charlies Cars, Bournemouth, bought many Albions from its first in 1928 to its last in 1960. The last batch included Harrington-bodied 1179CD, which is seen passing Lyndhurst, in the New Forest, in 1963. Charlies Cars were acquired by rival Shamrock & Rambler in 1963. The latter company passed to Hants & Dorset in 1966. This vehicle, along with the rest of the batch (1175-80CD), was bought by Morris, Pencoed, in either 1969 (1175-77CD) or 1971 (the remainder). 1179CD received accident damage in 1972 and was subsequently used as a source of spares. *Martin Llewellyn*

Left: The final offering to the coaching market from Albion was the front-engined Viking VK, which was introduced in 1963. Home sales were low, due no doubt to the competition from Bedford and Ford in this field, although it sold well overseas.

Perhaps the most stylish bodywork to be carried by a Viking was the Park Royal Royalist. Only six were built, four of which went to Red House, Coventry. One of these was KHP778E seen here in 1977 with Price (P+E TRavel), Aberbargoed, Mid-Glamorgan. New 10 years earlier, it had stayed with Red House until 1976. *John Jones*

Above: Although Atkinson had been building steam, and later diesel, lorries for many years, the company's first PSV chassis, the Alpha PM, did not appear until 1950. A lightweight underfloor-engined chassis, the Alpha PL, appeared in 1953, but few were sold. VS6440, seen in 1972 with A. & C. McLennan of Spittalfield, is a Gardner 5HLW-engined PL745H. Carrying a Duple body, the coach was new in 1955 and was acquired by McLennan from MacRae of Fortrose in 1968. *Stewart J. Brown*

Left: Jersey was a veritable haven for the old and unusual in the 1950s, although this Mann Egerton-bodied Austin CXD of Shamrock Tours was rather more modern than the Leyland Cubs, Lions and others still running around on the island at that time. The forward-control CXD was built between 1948 and 1955 and saw use mainly as coaches. Another popular model was the normal-control CXB, which was built to compete with the Bedford OB. Photographed at St Helier in August 1958, J16584 looks rather sad and could maybe do with a 'Crazy Nite' out. *Roger Holmes*

Above: The Barton fleet during the 1950s and 1960s had a high turnover of both new and secondhand vehicles and was a fascinating operator for the enthusiast. In addition to all the bus buying, a great deal of rebuilding and rebodying was carried out. Indeed the last rebuild took place in 1971 when a new Plaxton body was fitted to a rebuilt AEC Reliance chassis.

In 1953/54 some 17 Leyland Tiger PS1s were rebodied by Plaxton, whilst one was rebodied by Barton itself and classified BTS1. No 713, seen here at Nottingham in 1965, was formerly a Duple-bodied PS1 of 1947, but carries here a Plaxton C37F body. To the left is No 1023, a Harrington-bodied C41F Bedford SB5, one of 15 delivered in 1965, while in the background is No 772, a 1955 AEC Reliance/Duple C43F bought from Creamline (Bordon) in 1957. *Martin Llewellyn*

17

Above: In 1946, the Dartford coachbuilder J. C. Beadle constructed four chassisless single-deckers using reconditioned running units. Over the next 11 years, until the company closed in 1957, many integral vehicles were built, using a variety of engines (such as Bedford, Leyland and AEC). Following the takeover of Beadle by the Rootes Group in 1951, use was made of Commer components and, from 1954, the new Commer TS3 was incorporated.

Potteries took five of these integrals, Nos C7716-20, in 1956 with 41-seat bodies. The last of the batch is seen shortly before withdrawal in 1967; the whole batch was sold in 1968. *Mike A. Sutcliffe*

Right: Duple Vista-bodied OB was MAF446 in the fleet of Harvey, Mousehole. It is seen in 1967. *Mike A. Sutcliffe*

Left: A further Scottish OB was Scottish Motor Traction No C164, one of 13 delivered with SMT C29F bodywork in 1947 but rebodied by Burlingham to FC24F in 1953. The vehicles remained in service until 1961-62. All except one passed for further use to Highland Omnibuses and No C164 lasted until 1964. The coach - in its rebodied guise - is also seen in Portree, whilst on tour duties in 1959.
Roger Holmes

Right: The Bedford OL was a goods chassis that was also adaptable for passenger use. Available until 1953, it had a shorter wheelbase than the OB. The Isle of Man Road Services ran one of the type, KMN938, which dated from 1949 and carried an Armoury C29F body. This was one of three vehicles taken over with the stage service of Broadbent (Safeway Services) of Ramsey in 1950. KMN938 ended up with a Douglas dealer in 1966. *Mike A. Sutcliffe*

Left: The Little Bedford VAS appeared in 1961 - some 10 years after the demise of the OB - and was designed as a 29-seat coach. With its 16in wheels, it gained popularity with those operators wanting something smaller than the Bedford SB and, although other types were introduced by Bedford in later years, the VAS soldiered on until the end of Bedford production in 1986.

Bloomfield Coaches of London entered exam-ples of all four current Bedford models - a J2, SB, VAS and VAL - in the Ninth British Coach Rally at Brighton in April 1963, with at least the VAS also appearing at Blackpool in the National Coach Rally in the following May. The VAS entered was 8000MD, a VAS 1 with Plaxton Embassy C18F bodywork. The reduced seating, together with the destination as San Francisco, would suggest its use for longer distance touring work! *Michael Fowler*

Above: For operators wanting a small capacity vehicle, one of the Bedford commercial vehicle chassis could be converted. MacBrayne bought Bedford OLs in 1952 and C4/C5s with Duple bodies between 1958 and 1961. These included No 185, 609CYS, one of eight C5C1s delivered in 1961. It is pictured at Armadale Pier in 1965. *Martin Llewellyn*

23

Above: In 1950 Bedford announced the successor to the popular OB in the shape of the 33-seat SB. As it turned out, it was a case of one classic replacing another, as the SB continued in production until 1986 and the end of Bedford itself as a bus chassis builder - something of a record in PSV terms.

Until 1955, when production was transferred to a new plant in Dunstable, the SB was available only with a 17ft 2in wheelbase.

Eastern Counties took 18 SBO/Duple Vega C37F coaches in 1954. These were fitted with Perkins oil engines. No BV 854 (PPW854) is seen in Norwich in 1965, the year before the whole batch was withdrawn. The buses in the background are all Bristols, the traditional choice, of course, for a BTC operator. However, a few other Bedfords were bought - OBs in 1949-50 and VAMs in 1967-68, for example.
Martin Llewellyn

Right: The last petrol-engined coaches to be bought by Alexanders were a batch of 10 Bedford SBGs (the 'G' stood for gasoline) with Burlingham C35F bodies delivered in 1956. When Alexanders was split in 1961, HMS226, as illustrated here in Aberdeen, passed to Alexander (Northern). Withdrawn in 1968, HMS226, along with sister coach HMS227, was sold to Leask of Lerwick. *Stewart J. Brown*

Left: Another Scottish Bus Group member to operate the SB was Highland Omnibuses, although these were all various secondhand acquisitions. In December 1967 Highland bought seven coaches from Alexander (Northern), all of which had originated with Simpson's Motors (Rosehearty). These consisted of four Ford 570Es and three Bedford SB1s. All eight carried Duple C49F bodies and were new in 1960.

Illustrated is No CD25 at Portree in 1970. Despite carrying the blue and grey coach livery, the vehicle had in fact been fitted for one-man operation and had had its seating reduced to 39. In addition, front destination boxes had been fitted. The conversion took place before entering service. The whole batch was withdrawn in 1973 and all three SBs passed to Peace, Kirkwall, where they lasted a further four years.
Chris Lodington

Above: The Bedford SBs illustrated so far are perhaps atypical of the breed in so much as the type has always been seen as the workhorse of the smaller independent, whether for school or contract duties, weekly trips to market or to the seaside in the summer. Representing a more typical SB is this 1963 example with Duple Bella Vega C41F bodywork of Potter, Skewen (near Swansea). It was new to Whittle, Highley.
John Jones

Left: Of all the coaches illustrated in this book, the six-wheel, twin-steer Bedford VAL is the most distinctive and easy to identify. This revolutionary, if not unique, chassis (Leyland built the twin-steer Gnu before the war, for instance) was launched in 1962 with the extra axle and small 16in wheels offering more stability for the front-mounted engine.

These impressive vehicles sold well over the following 10 years, although only in small numbers to the larger operators. 5188RU is a Leyland 0.400-engined VAL 14, which was new in June 1963 to Excelsior Coaches, Bournemouth, as one of a pair - the other was 3711RU - that carried Plaxton C52F bodies. After three years of tour and excursion work it was sold to West Wight Motor Bus Co, Totland Bay, Isle of Wight, where it remained until acquisition by Calvary coaches, Washington, in 1987. The veteran makes a fine sight in the summer sunshine in Sunderland later that year. *Kevin Lane*

Above: The large coach operator Wallace Arnold was an enthusiastic VAL customer as the type proved popular on extended tour work. As a result both the VAL 14 and the later VAL 70 - the latter using the Bedford 460 engine - were taken into stock. BNW620C is a VAL 14 and carries a Plaxton Panorama body of a more modern design to that of 5188RU illustrated earlier. The number of seats was reduced to 49 in view of the vehicle's duties on long-distance tour work. It is seen when new in 1965 in Coventry. It lasted with Wallace Arnold until 1970. By the end of the 1970s, BNW620C had passed, along with many other VALs, into non-PSV use. *Martin Llewellyn*

Left: The Bedford VAL with Duple Vega Major bodywork was immortalised by Dinky Toys in the company's model of the type (how I wish I still had mine!). 45SAU was delivered in 1963 to Skill, Nottingham, as one of eight VAL 14s supplied to the fleet with Duple bodies that year. Two Plaxton-bodied examples followed in 1964. The coach is seen swinging out of Huntington Street bus station, Nottingham, bound for Bilsthorpe Colliery in December 1965. *Martin Llewellyn*

Upper right: Duple and Plaxton were responsible for the majority of the coach bodywork fitted to the Bedford VAL, although by the 1960s there was little other choice anyway. Yeates of Loughborough bodied just 11, and only four of these were constructed as coaches. Two of the quartet went to Gibson of Barlestone in 1963: 179CNR in May and 407EAY followed in October. The clean lines of the Fiesta bodywork is shown to good effect on the latter when photographed in Leicester. *Mike A. Sutcliffe*

Lower right: Harrington designed its 'Legionnaire' for the VAL, with 42 being built in 1964-65. Yelloway of Rochdale took four out of a fleet of 19 VALs operated; the remainder carried Plaxton bodies. One of the Harrington-bodied VALs, CDK410C, is pictured during September 1965 in Leicester, Blackpool-bound, when only a few months old. *Mike A. Sutcliffe*

Left: The 45-seat VAM appeared in 1965 and this was replaced by the YRQ from 1970. The front-engined VAM sold mainly to independents, although it was also acquired by larger operators such as Southern Vectis (a company which had a long Bedford tradition), Crosville, Highland and Trent, albeit only in small numbers.

Illustrated here, however, is a municipal VAM 70. BJH128F was new to North Star Coaches, Stevenage, in 1968 and was acquired by Barrow Corporation for use by the council's Social Services Department. It remained with the fleet until 1978. Barrow's coach operation increased 10-fold in 1973 when Hadwin of Ulverston was taken over. This brought eight Bedford VAL 70s and two VAL 14s into the fleet. *John Jones*

Above: Bedford had a loyal following, with some operators buying little else and just updating their fleet from the available range from the company. One such operator was Barnes Coaches of Clacton-on-Sea, who bought only Bedfords for many years. Four of the fleet were lined up in the yard in Pier Avenue on a Sunday in June 1980: VAL 70s VNK911J and LAW897F; SB5 MMC302C; and, VAM 70 HXG303F. All four were fitted with Duple bodies. Barnes Coaches ceased operation in September 1987. *Kevin Lane*

Left: The Birmingham & Midland Motor Omnibus Co - Midland Red for short - built its own buses from 1923, designs which were in many cases well ahead of their time.

The first postwar coaches were designated C1 and consisted of a batch of 45 underfloor-engined 30-seaters with Duple bodies, which were delivered in 1949 - well ahead of similar offerings from AEC, Bristol and others. The vehicles had a long life with Midland Red, No 3318 being pictured at Leicester in 1965. *Mike A. Sutcliffe*

Above: The coaching fleet steadily developed through the 1950s and in 1958 the prototype C5 appeared. This was based on the S14/S15 bus which was then in service. With the opening of the M1 motorway in 1959 there followed the turbo-charged CM5 and toilet-fitted CM5T. The increase in maximum legal length for buses from 1961 saw the arrival of the 36ft motorway coach - the CM6T - with the first production models entering service in 1965.

CM5 No 4833 of 1961 waits at Worcester before returning to Birmingham in 1964. *Martin Llewellyn*

Above: CM6T No 5660, seen in later years, is London-bound at Coventry in 1972, the year in which the last of the C5 type was withdrawn along with the first of the CM6Ts. A lack of spare parts saw all of these vehicles off the road by 1974, leaving Leyland Leopards on the motorway services in replacement. *John Jones*

Right: The first postwar Bristol single-deck and coach chassis was the L, introduced in 1946, which was an updated version of the 1937 model. This was replaced by the longer LL from 1950 when a 30ft maximum length vehicle was permitted. In addition an 8ft wide version was produced from 1951 to 1954; previously these models had been manufactured for export only.

Royal Blue was soon to restart its express routes to the West Country, after the cessation of hostilities, in 1946, but was not able to re-equip its fleet until 1948-49 when 45 Bristol Ls with Beadle bodywork were delivered.

No 1270, a 1951 Duple-bodied L6B, is seen at Victoria preparing to leave for Weymouth in September 1962. By this date the vehicle was looking remarkably dated. *Iain MacGregor*

Above: Although of the same vintage as the previous vehicle illustrated, Western National No 1310, an LWL6B, carries an ECW FC37F body of much more modern appearance. Seen in Perranporth in 1961, the vehicle lasted another four years until withdrawal and subsequent export to Cyprus. *Roger Holmes*

Right: Bristol's first underfloor-engined chassis was the LS. The 'LS' stood for 'Light Saloon' and the type was first introduced in 1950.

Southern Vectis No 303 was the first LS delivered to the company and was one of four delivered in 1952. It was fitted with ECW C39F bodywork and was photographed climbing out of Ventnor in 1963. No 303 was converted to one-man operation by Strachans in 1966 and received green and cream livery at the same time. It was withdrawn in that condition in 1970. *Martin Llewellyn*

Above: In 1957, the Bristol MW ('Medium Weight') superseded the LS as the standard BTC coach and single-deck bus chassis. The early ECW body style was rather conservative in comparison with many of its contemporaries, but was attractive nevertheless. United No BUE530, 530LHN, takes a break outside Chambers cafe, Doncaster, whilst working a northbound Thames-Tees-Tyne service in 1963. The coaches on this service carried a green and cream livery originated by Orange Bros, one of the main instigators of the London-northeast service, which had been taken over by United in 1934. *Roger Holmes*

Right: The later style of ECW coach body fitted to the MW was still rather conservative, but did look less like a service bus than its predecessor. Royal Blue No 1383 was new in 1962 and is seen empty in Southampton in August 1974. *Mike Greenwood*

Above: For something a little smaller than the MW, Bristol developed the SU ('Small Underfloor-engined'), which was built between 1960 and 1966. Of 181 built, only 38 were coaches. Two of the 38 went to United Welsh whilst the remainder went to Southern/Western National. Western National No 416 is a 1961 SUL4A with ECW C33F bodywork. *Mike A. Sutcliffe*

Right: The first generation of rear-engined single-deckers were not the most reliable of buses, but the Bristol RE, launched in 1962, was undoubtedly the best of the bunch. It saw widespread use as a coach among THC operators and later with National Express until ousted by the Leyland Leopard. Only a handful were ever bought new by independents, despite the ability of Bristol to sell on the open market from 1965. The last coaches were the dozen delivered in 1975, the last numbered chassis ironically going to an independent - Davies of Halewood.

United Counties No 271 is a typical 36ft RELH with the earlier style of ECW coach body seating 47. It is pictured at a damp Nottingham in August 1971 on the London service and provides a contrast with a Royal Blue Bristol MW on Associated Motorways duties. *G. H. F. Atkins*

Above: A shorter chassis, the RESH, was also available for coach applications, but only 11 of these 33ft-long vehicles were built. All but two carried Duple Northern bodywork, which added to their rarity.

Southern Vectis bought only two Bristol RE coaches, preferring to stick with Bedford. One was an ECW-bodied RELH, whilst the other was No 301, a Duple Northern-bodied RESH6G that was new in 1968. Although it had been painted into National white livery in 1974, it was subsequently repainted in Tilling green and cream (much nicer!) as seen here at Old Shanklin in 1978. *G. R. Mills*

Right: Bristol launched the mid-engined LH in 1967 and thus this was the company's first chassis to be available immediately on the open market. Three variations were available, which could accommodate 26ft, 30ft or 36ft bodywork. The type sold well in both the bus and coach market; the latter particularly to the private sector.

Grey Green Coaches was an early customer, buying one LH and three LHLs, all with Plaxton bodies, in 1969. The LH, AGH590G, is seen picking up at Lexden, Colchester, *en route* for London. *G. R. Mills*

Above: The first three longer LHLs went to Golden Miller of Feltham in 1968 as TMT763-765F. All three carried Plaxton C53F bodies. By 1973 TMT763F had passed to Kenfig Motors, Glamorgan, where it looks well turned out. All of the LHLs went to independent operators, whilst several were even bodied as pantechnicons for the manufacturers of Silver Cross prams. *John Jones*

Right: Several operators of the Bristol Lodekka specified coach seating for use on limited stop services. Eastern National ran 15 FLFs on services between London and the Southend area. The first were Nos 1608-10, which were delivered in 1962 and fitted with ECW CH37/18F bodies. No 1610 is pictured when only a year old on layover from the X10 Southend service at Victoria in the company of 1960 Bristol MW/ECW No 508, also of Eastern National, and Aldershot & District AEC Reliance/Park Royal No 472. *G. H. F. Atkins*

Left: The Crossley SD42 was a popular choice for a number of postwar coach operators. Gash of Newark, normally a confirmed Daimler operator, bought one in 1948. This was KAL551, which was fitted with a Yeates C35F body. After 15 years with Gash, it passed to non-PSV use in 1963 and later into preservation. Crossley was taken over by AEC in 1948. This led to the gradual winding down of production, with the last SD42 appearing in 1953. *Arnold Richardson/Photobus*

Below left: After the war Daimler introduced its Victory range, of which the most popular choice for coach work was the CVD6 which appeared in 1946. South Yorkshire independent operator Leon of Finningley operated this Plaxton-bodied CVD6 from 1948 until 1963. The vehicle was acquired second-hand from Robinson of Great Harwood when it was only a year old. It is seen in Doncaster in the early 1960s. *Arnold Richardson/Photobus*

Right: It was fashionable in the 1950s to fit full-fronted bodywork on to previously half-cab vehicles to give a more modern appearance. This CVD6 of E. A. Hart Ltd (Beehive) of Doncaster received a new Duple body in 1955. The new body considerably improved its image. *Roger Holmes*

Left: Daimler entered the underfloor-engined market in 1951 with the Freeline, which was the company's first chassis to receive a name. It did not, however, enjoy the success of some of its contemporaries produced by AEC and Leyland. Less than 100 were sold in this country, although it did well abroad. Production ceased in 1964. RFM700 carried a Duple C41C body when photographed in 1970 parked up at Drefach, Carmarthenshire. At this time it was operated by Williams, Gwaun-cae-Gurwen, but was originally new to Taylor of Chester in 1953. *John Jones*

Above: We pass now on to the Daimler Roadliner, which was a Cummins-powered rear-engined chassis that found few buyers. It was launched at the 1964 Commercial Motor Show at a time when single-deckers were coming into fashion, but the model's unreliability was its downfall, and could not even be saved by a change to Perkins or AEC running units. Most sales, both at home and abroad, were as buses, with only 54 being constructed as coaches for use in this country. Out of this total 38 went to Black & White Motorways, who specified Perkins engines for the last 20.

Central Coaches, Walsall, bought two - MDH211/212E with Plaxton C51F bodies - in 1967. Both were Cummins-engined SRC6s. The former is seen at Silver End, Essex, prior to sale to Clark, London SE9, in February 1978. *G. R. Mills*

Left: The Dennis Lancet 3 appeared in 1945 and was a popular choice as a coach chassis, particularly with independents. BCB478, fitted with a Duple body, was new in 1948 to Valiant Cronshaw. It later passed to Smith of Reading as seen here. *Alistair Douglas/Photobus*

Above: As with the other major manufacturers of the period, Dennis unveiled its first underfloor-engined chassis, the Dominant, in 1950. This was rather a false start as it turned out; the advanced design, incorporating semi-automatic transmission and turbo-charging, was perhaps ahead of its time. The more conventional Lancet UF appeared in 1952, meeting with moderate success.

The first large order for the Lancet UF was for 30 with Duple bodies delivered to East Kent in 1954, HJG3-32. Although the batch had long lives with the company - they were withdrawn between 1968 and 1971 - no more of the type were acquired. East Kent's coaching requirements were met mainly in the future by AEC Reliances. HJG4, seen at Dover in 1962, lasted until 1969 before passing into non-PSV use. *Roger Holmes*

Above: Foden did not build its first buses until 1945, having been a manufacturer of steam and diesel lorries for many years. The chassis used as the basis for single-deckers was the front-engined PVSC, which was available for 10 years from 1946. A rear-engined coach, the PVR, was also offered from 1950, but it met with little success.

Foden PVSC CAG802 of Dodds, Troon, dates from 1948 and carries a Plaxton body with later front end. It was photographed in 1962, looking in remarkably good condition considering its age. *Iain MacGregor*

Right: Looking rather more like a Foden, due to the distinctive concealed radiator design (which was also used on the company's lorries at this time), is this PVSC6 with Burlingham FC35F bodywork operated by Strachan's Deeside Omnibus Service of Ballater. JWU751 was new to Larratt Pepper of Thurnscoe in 1950, passing to Strachan in 1959. Strachan was taken over by Alexander (Northern) in 1965 and the Foden was sold for scrap later that year. *Arnold Richardson/Photobus*

Left: Ford began production of the Thames Trader in 1958 in direct competition with the highly successful Bedford SB. A derivative of the company's goods chassis of the same name, the forward-control, front-engined design won many friends and became a popular choice of independent operators in the early 1960s.

Hanson, Huddersfield, was a good customer and four of the company's 1964 intake were Fords, including No 387, BCX490B, which was fitted with Duple Northern 'Firefly' bodywork. The coach is seen at Chester in 1967. The Hanson business was taken over by West Yorkshire PTE in 1974 and formed the basis of the PTE's coach operation. This was sold in 1979 to Abbeyways of Halifax. *Martin Llewellyn*

Above: The Thames Trader was replaced by the R-series in 1965. This sold well in Scotland and is exemplified by Highland No T10, DST440D seen in Oban in 1974. This was an R226 with Plaxton Panorama C52F bodywork. New in 1966, the coach was withdrawn and sold in 1980. The R-series saw considerable sales during the 1970s, including a number to the National Bus Company, but production ceased in 1985. *G. R. Mills*

Left: The little Guy Otter was designed for bodywork accommodating up to 30 seats - ie what today would be considered a midi-bus or coach. The rather unassuming vehicle illustrated here is, in fact, a rare, if not unique, beast. No 2 in the fleet of Hulley's, Baslow, was NTB403, a Guy Otter with Alexander FC29F bodywork. Exhibited at the 1950 Commercial Motor Show, it passed initially to Monk, Leigh, and then to Sutcliffe, Burnley, and MacFarlane, Balloch, before reaching Hulley's in 1959. It remained with Hulley's until 1971 when it was bought for preservation. Its future was not secure, however, as it was sold for scrap in 1984. This 1960s view is in Chesterfield, probably on the Tideswell route, which was its regular haunt. *Arnold Richardson/Photobus*

Above: Another Guy rarity was the front engined Warrior, which was introduced in 1956 to satisfy export customers who wanted a simple, rugged machine. Only two saw service in this country: demonstrator VDA32, which was fitted with a Willowbrook bus body, and JBV234, illustrated here. New in 1957 to Ribblesdale, Batty Holt, Blackburn, the coach had a Plaxton C39F body. It is seen here in Tomintoul on tour work in August 1959. Withdrawn after six years with Holt, JBV234 eventually ended up with Hulley's, Baslow, in 1968 (obviously a haven for interesting Guys!), where it lasted four years. *Roger Holmes*

Above: Guy unveiled its first underfloor-engined chassis at the 1950 Commercial Motor Show. Designated the Arab UF, it incorporated a Gardner HLW engine - a definite plus point over some of its competitors. It was still a heavy chassis and further development led to the lighter Arab LUF in 1952. Both the UF and LUF ceased production in 1959.

Red & White bought its first underfloor-engined coaches, Leyland Royal Tigers, in 1951, although Guy Arabs UFs with Duple bodywork were bought alongside more Leylands in 1952. However, with Red & White now under BTC control, all new vehicles were of Bristol manufacture from 1954 until the first Leyland Nationals in 1974. The company's coaching

requirements were thus fulfilled by the LS, MW and RELH models. However, a number of secondhand vehicles were acquired during this period, including a quartet of Guy Arab UF coaches with Gardner 6HLW engines which came from Thames Valley in 1960. New in 1952, the four carried Lydney C41C bodies completed by Bristol after Lydney had closed downand had worked on South Midland services. South Midland had been part of the old Red & White empire, but had passed to Thames Valley after Red & White had become British Transport Commission-owned in 1950.

SFC502 was initially numbered U1652 by Red & White but was downgraded to No DS1652 in 1962. It is seen in this latter state. It passed to a

Lancashire dealer on withdrawal in 1965 and ended up in non-PSV use. *Mike A. Sutcliffe*

Right: With the war over, Leyland began production of the Tiger PS range of traditional front-engined arrangement. Red & White acquired 20 PS1/1s with assorted bodywork when the Griffin Motor Co, Brynmawr, was taken over in 1953. These included EU9278, which was new in 1949 and was fitted with Duple 'A' C35F bodywork. The vehicle was downgraded to bus work and numbered S3048. It is seen in the company of two other former Griffin vehicles: Utility Guy Arab IIs with Bristol L27/28R bodywork Nos L644/744. The trio was withdrawn in 1964. *Mike A. Sutcliffe*

Above: As we have seen, full-front bodywork was popular in the early 1950s in order to disguise the half-cab layout and to give a more modern appearance. This Leyland PS1/1 carries Burlingham FC33F bodywork and was new in 1950 to Dimbleby of Ashover. It remained with the operator for 17 years. Alongside is HTC863, a Bedford OB with less convincing full-fronted bodywork, this time by Plaxton. The view dates from September 1965; Dimbleby ceased operations in 1980. *Michael Fowler*

Right: Although there had been underfloor-engined Leylands before the war, it was not until 1949 that the integral MCW/Olympic was launched. The type was not a success as operators preferred to choose their own bodywork. This resulted in the Royal Tiger being launched in 1950. The Royal Tiger remained in production for only four years, being superseded by the lighter Tiger Cub.

East Kent added eight Royal Tigers to its fleet - in 1951 (three), 1952 (three), and, 1953 (two) -

before turning back to Dennis for its coach deliveries in 1954. FFN451, with Park Royal C37C bodywork, was the first of the three delivered in 1951. It was photographed at Dover in the company of EFN595, a Park Royal-bodied Dennis Lancet J3, which was only a year older. FFN451 was withdrawn in 1967, while the Dennis was downgraded to bus work in 1961, surviving in that role for another three years. Both vehicles were still in non-PSV use until the mid-1970s at least. *Martin Llewellyn*

Left: Yorkshire Traction bought the Royal Tiger in 1951 and 1952 largely for bus use. The exceptions were six PSU1/15s delivered in 1951, which carried stylish Windover C37C bodies. The same bodybuilder had also been patronised for coaches on Dennis Lancet chassis in 1949 and Leyland Tiger PS2/3s in 1950. No 924 (DHE562) was photographed on home ground in Doncaster in 1959, three years before withdrawal, on its way to the races. *Roger Holmes*

Below left: Leyland offered its own design of bus and coach bodies for the Royal Tiger from 1950. While the bus body was quite 'boxy', that for the coach was more stylish, with a distinctive windscreen arrangement as seen on Hylton & Dawson RPG807, a 1952 PSU1/15 with Leyland C41C body. New to Hunt, Ottershaw, it passed to Hylton & Dawson in 1957, being sold to an Irish operator in 1966. The coach is pictured pulling out of the much-photographed bus station at Leicester St Margarets in 1964. *Mike A. Sutcliffe*

Right: A number of vehicles passed to Mexborough & Swinton from fellow BET operator, Southdown, during the early 1960s, including this late Royal Tiger PSU1/16 with Harrington C26C body. This vehicle was transferred to the Yorkshire operator in 1965, becoming No 103 in the M&S fleet. Seen at Rawmarsh in 1966, it lasted in service until withdrawal in 1968. *Michael Fowler*

Above: The well-known County Durham independent operator, OK Motor Services, has only operated one Leyland Royal Tiger - PSU1/15 EJR791. New to Hunter, Seaton Delaval in 1952, it originally carried a Burlingham C41C body, but received a Plaxton body in 1966. The vehicle passed to OK in 1977 and thence to Lockey, West Auckland, in 1985. OK had acquired Lockey in 1983, running it as a subsidiary company until final absorption in 1985. Thus, EJR791 returned to OK for another two years prior to final withdrawal in 1987. The photograph shows the vehicle at St Helens, Auckland, in 1981. *John Jones*

Right: 1952 saw the appearance of the Leyland Tiger Cub PSUC1, which was lighter than the Royal Tiger. The Tiger Cub proved to be quite a success, continuing in production for 17 years.

Silcox, Pembroke Dock, bought its first Tiger Cub in 1954, Burlingham-bodied SDE400, and subsequently owned a number, although these were mostly acquired secondhand. SDE400 lasted some 20 years with Silcox before being scrapped. Sneaking into the picture on the left is Leyland Leopard PSU3/3R 734PDE, fitted with a Willowbrook C51F body, which was new in 1962 - the first 36ft-long vehicle in the Silcox fleet. *John Jones*

Left: Lancashire United took eight Tiger Cub PSUC1/2s into stock in 1960. These were fitted with unusual Northern Counties C41F bodies similar in design to the Duple dual-purpose bodies, as illustrated here by No 35 128MTE at Blackpool in June 1966. *Mike A. Sutcliffe*

Below left: Truly a class, the Leyland Leopard, which was initially a more powerful Tiger Cub, first appeared in 1959. The coach version was designated the L2.

Llynfi Motor Services, Maesteg, No 31 was a 1961 L1 with Plaxton Panorama C41C bodywork. It was originally new to Wallace Arnold as one of a batch of eight (9901-8UG). Withdrawn in 1969, 9905UG passed through several new owners before reaching Maesteg in 1976. However, it lasted only a year before withdrawal and scrapping in 1978. The photograph was taken at Sophia Gardens, Cardiff, in June 1976. *John Jones*

Right: With new regulations came the 36ft-long bus and a new Leopard was produced to take advantage of the change - the PSU3 - introduced in 1961. The redesigned L range became the PSU4 in 1964.

The first coach service from Scotland to London began in 1928 and took two days to complete. By 1962 this was down to 13hr and 40min and, in the following year, Western SMT placed its first 36ft coaches on to the service in the shape of 13 Alexander C38F-bodied PSU3 Leopards, Nos KL1824-36, which featured toilets, forced air ventilation and reclining seats. No KL1831 is seen in June 1963 about to pick up for a night Glasgow-London run. *Iain MacGregor*

Above: Unique to Southdown were 20 Leopard PSU3s with Weymann Castilian bodywork delivered in 1962-63. The first five (Nos 1155-59) had seven bay windows, whereas the rest were tidied up and had just four and a half bays. One of the earlier batch, No 1155, is pictured on hire to Eastern National pausing for a refreshment stop at the Duke of Wellington, Hatfield Peverel, whilst working between Clacton and London, Victoria, in February 1967. *G. R. Mills*

Right: In September 1969, United Counties acquired the London-Rushden service of Birch Bros, London, together with 12 Leopard PSU3s which were fitted with a variety of bodywork supplied by Willowbrook, Park Royal and Marshall. Loading at Bedford shortly after the take-over is No 191, 91FXD, a 1963 PSU3/3R with Park Royal C49F bodywork. The vehicle still carries Birch livery, but has received a red and black Rushden garage/fleetnumber plate and carries the new owner's name in the windscreen. Renumbered 238 in 1974, 91FXD was withdrawn and sold later that year. It was subsequently re-engined and fitted with a Plaxton body to emerge as PJP275R in 1977. *Mike A. Sutcliffe*

Left: Beetwell Street, Chesterfield, is pictured in September 1970 as two Leyland Leopard PSU3s pause on their journey from Coventry to Newcastle. Nearest the camera is Yorkshire Traction No 16, a 1967 PSU3/4R with Plaxton Panorama C49F bodywork, while behind is Hebble No 411, an Alexander C45F-bodied PSU3/4RT recently transferred from the Yorkshire Woollen District fleet. *G. H. F. Atkins*

Above: With the Leyland Atlantean the best-selling rear-engined double-deck bus of the 1960s, thoughts turned to a rear-engined single-decker. The resulting Panther and Panther Cub left a lot to be desired, despite the strong influence of the successful Leopard and Tiger Cub in their design. Most were bodied as buses, although a few Panthers were built as coaches.

Pam's Coaches, Enderby, Leicestershire, was running this ex-Seamarks, Luton, Plaxton-bodied Panther PSUR1/2/2 SXD473F on hire to National Travel when seen at Bournemouth in June 1975. *John Jones*

Left: East Yorkshire ran both Panthers and Panther Cubs as buses and also had Panthers as dual-purpose vehicles and full coaches. Among the last-mentioned were two PSUR1/2Rs with unusual Metro-Cammell C44F bodies, Nos 823/24 (JRH323/24E). The former is seen outside Anlaby Road, Hull, garage in 1972. New in 1967, No 823 lasted nine years in the fleet before withdrawal and scrapping. *Michael Fowler*

Upper right: In 1948 Ribble placed the first of its 'White Lady' double-deckers into service. The Leyland PD1/3 chassis carried Burlingham 49-seat coach bodies, which were very sumptuous for the day in featuring air conditioning and with the upper saloon being provided with a twin sliding sunshine roof. This initial batch of 30 was followed by 20 similar bodies built by East Lancs on PD2/3 chassis in 1950-51. The new Leyland Atlanteans were used for the next generation of double-deck coaches from 1959. The 'Gay Hostesses' featured a toilet and refreshment facilities, the latter being provided by an onboard hostess or steward. The vehicles were distributed variously among the Ribble, Standerwick and Scout fleets.

Standerwick No 28, a Weymann-bodied Atlantean PDR1/1, is pictured London-bound loading at Pool Meadow, Coventry, in September 1962. *Iain MacGregor*

Lower right: The postwar model offered by Maudslay was the normal-control Marathon 3, which was almost identical to the AEC Regal III. Maudslay, along with Crossley, was taken over by AEC in 1948, with production ceasing two years later.

Smith, Grantown-on-Spey, operated three Maudslays, which had been acquired from MacBraynes in 1962. FUS997 dates from 1948 and carries a Park Royal C35F body, in common with the other two. The battered looking vehicle next to it is GSC457, a Burlingham-bodied AEC Regal III, which was new to Scottish Omnibuses in 1949, passing to Highland Omnibuses in 1962 and thence to Smith in 1963. Smith was taken over by Highland in 1966, but neither coach survived to be taken over at that time. *Alistair Douglas/Photobus*

Left: The immediate postwar period, up until the early 1950s, was a boom time for the British bodybuilder. However, the general decline throughout the bus industry saw all but a handful disappear before the end of the decade and by the end of the 1960s the only real choice, if you wanted to buy a coach body, was between Duple and Plaxton. Similarly, following closures and take-overs, there was not much choice if you wanted a British chassis as well, although the likes of AEC, Bedford, Ford and Leyland were still in the market.

The dissatisfaction of some operators with the standard of quality and service available in the late 1960s saw them looking to Europe for alternatives and the rest, as they say, is history.

Continental coaches had been appearing at early British Coach rallies for a number of years; a rather severe-looking Van-Hool/Fiat, for example, competed at the 1960 show, but the first actual demonstrator was a Mercedes Benz O.302, OLH302E. This was an 11m, 43-seat integral vehicle and is pictured at the 1967 Brighton Rally. It was tested by the Scottish Bus Group and various BET and THC operators, before being sold to Whiteford of Lanark. *G. R. Mills*

Lower left: Coach operators Rowe's Garages of Dobwalls, Cornwall, built a prototype chassis in 1953. This was followed by the production of four further PSVs and over 100 commercial vehicles by a new company, M. G. Rowe (Motors) Doublebois Ltd, until it was wound up in 1963.

The initial chassis, the Hillmaster, appeared as a front-engined lightweight model, which was powered by a Meadows 4DC engine and carried a Whitson C38C body. It only ever ran on trade plates and was subsequently rebuilt as an underfloor-engined vehicle; the original bodywork was altered to fit, losing a seat in the process. Thus, TAF587 was shown at the 1954 Commercial Motor Show at Earls Court. Rowe's operated the coach until 1973, after which it was sold to a riding school who, unfortunately, scrapped it in 1977. TAF587 is seen here looking very smart in August 1967.

The four other Rowe Hillmasters were all buses and one, Reading-bodied WRL16, has been preserved. *Mike A. Sutcliffe*

Above: Motor Traction Ltd, of Addington (Surrey), trading as Rutland, built a prototype rear-engined chassis, the Clipper, in 1953. In 1954/55 two chassis were bodied as coaches.

One of the duo, TKE741, seen here in the yard of dealers W. S. Yeates of Loughborough in 1959, had operated for Spiers of Henley. It carries a Whitson body. *Mike A. Sutcliffe*

Left: Truck builders Seddon began production of its first passenger chassis, the Mark IV, in 1948 and continued to build various front, mid and rear-engined models from the 1950s right through until the 1980s - latterly with the successful Pennine VII, which was developed for the Scottish Bus Group. Illustrated, however, is a rare bird, the Seddon XIX. Only 30 of these underfloor-engined chassis were built. All were fitted with AEC six-cylinder 7.7 litre engines and 29 went for export. The exception was VHO200, which was new to Liss & District in 1959. It carried a Harrington Wayfarer body. In 1961 it passed to its current owners, Thornes of Bubwith. It was, however, withdrawn and stored in 1971. It has since been restored to full PSV order by Thornes and is seen in this condition in 1993 whilst working the York city shuttle between the National Railway Museum and the Barbican via the city centre.
Martin Higginson

Above: The postwar Tilling-Stevens was an elusive machine, with few being built, although a number of models were offered. These included the Express K series and the lightweight Express Mk II. One of the latter, operated by N&S, Leicester, is pictured. This particular operator ran a number of this type. The front-engined model was no doubt unsuccessful as a result of the triumph of the all-conquering Bedford SB. Production ceased in 1953 after only three years. This particular vehicle, KAY900, fitted with Plaxton C37F bodywork, was first registered in 1954 and so may have been bought as a chassis only, as a number were advertised for sale by a dealer at around this time. This picture was taken in July 1963, so presumably N&S got a fair amount of work out of the coach.
Mike A. Sutcliffe